GIBBONS

MONKEY DISCOVERY LIBRARY

Lynn M. Stone

Rourke Corporation, Inc.
Vero Beach, Florida 32964

PHOTO CREDITS

All photos © Lynn M. Stone

ACKNOWLEDGEMENTS

The author thanks the following for photographic assistance:
Lowry Park Zoological Garden, Tampa, Fla.; Miami Metrozoo, Fla.

LIBRARY OF CONGRESS
Library of Congress Cataloging-in-Publication Data
Stone, Lynn M.
  Gibbons / by Lynn M. Stone.

  p. cm. — (Monkey discovery library)
  Summary: An introduction to the intelligent, agile acrobat of
the Asian forests.
  ISBN 0-86593-062-7
  1. Gibbon—Juvenile literature. [1. Gibbon.] I. Title.
II. Series: Stone, Lynn M. Monkey discovery library.
QL737.P96S773   1990
599.88'2—dc20   90-32340
                                 CIP
                                 AC

Gibbon

TABLE OF CONTENTS

THE GIBBON

Gibbons are the circus **acrobats** of the Asian forests, and gibbons perform their magic without a net.

No animal swings more skillfully from tree to tree than these large monkeys. In fact, the gibbons' scientific name, *Hylobates,* means "dweller of the trees."

There are nine kinds, or **species,** of gibbons. None of the species is nearly as big as the great apes—gorillas, chimpanzees, and orangutans. But they're big enough to be known as "lesser" apes.

Siamang Gibbon

THE GIBBON'S COUSINS

Although they lack great ape size, gibbons are closely related to great apes. They have teeth much like the apes. And, like their larger cousins, gibbons have no tail.

Like all members of the monkey kingdom, gibbons are intelligient. They also have monkey fingers and toes that grasp, or hold.

The gibbons' long arms and fingers help them move easily from tree to tree.

Orangutan

HOW THEY LOOK

A gibbon may be gray, black, brown, or cream colored. The color of gibbon fur depends somewhat upon the species of gibbon.

The largest of the gibbons is the siamang. A siamang is mostly black, but it has a pink or gray pouch of skin on its throat. The pouch fills with air like a balloon when the siamang calls, which it often does.

Gibbons are long and slim. They range in size from about nine pounds to nearly 30.

White-handed Gibbon

WHERE THEY LIVE

Gibbons live only in southeast Asia. They live in such countries as Thailand, Indonesia, Java, and Cambodia.

Forests are their homes, or **habitats.** Since gibbons spend most of their time in trees, scientists call them **arboreal.** Because they are arboreal, gibbons must have forest homes.

Most gibbons live in low, wet forests. The siamang, however, can also be found in forests on mountainsides.

Siamang Gibbon

Siamang Gibbon

White-handed Gibbons

HOW THEY LIVE

Gibbons spend their lives in the treetops. They feed in trees and sleep each night in trees.

Gibbons aren't timid about heights. They regularly use branches 90 feet above the jungle floor.

Many monkeys live in large groups. Gibbons live in small, one-family groups.

Each family of gibbons has an area in which it lives called a **territory.** Gibbons let other gibbons know where their territories are by calling loudly.

If a gibbon enters another's territory, the "owner" will shriek and chase the visitor away.

THE GIBBON'S BABIES

A female gibbon usually has one baby every three or four years. A mother gibbon needs her hands and feet for climbing and swinging. Her baby uses its hands and feet to hang onto mom, leaving her limbs free.

By age of five or six, a young gibbon leaves its parents. It may spend several years finding a mate of its own.

Captive gibbons have lived for as long as 34 years.

White-handed Gibbons

PREDATOR AND PREY

Gibbons are rarely attacked by **predators,** or hunting animals. Hunters like tigers and leopards can not chase gibbons 90 feet up a tree.

Gibbons themselves are mainly plant eaters. The siamang, for example, lives mostly on leaves and fruit. Other gibbons, such as the white-handed gibbon, eat fruit almost always.

Gibbons also eat flowers and buds.

The only animals that gibbons are likely to eat— their **prey**—are insects.

Siamang Eating

THE GIBBON AND PEOPLE

People who handle captive gibbons work carefully. Gibbons may be gentle, but they can also become angry quickly.

People enjoy watching gibbons. They are amazing acrobats, and they have many loud, strange calls.

Like all monkeys, gibbons are often humanlike in the way that they behave. But having some humanlike ways hasn't helped gibbons in the wild. Gibbons have been hunted too often, and much of their habitat has been destroyed.

Siamang Gibbon

THE GIBBON'S FUTURE

People don't destroy the forests to destroy gibbons. They cut forests because they need farmland and wood.

As the number of humans in Southeast Asia grows, the forest shrinks.

All gibbons are **endangered** species. They are in danger of becoming **extinct,** which means disappearing forever.

The silvery gibbon of Java and the capped gibbon of Thailand and Cambodia are in the greatest danger.

Hopefully, the nations of Southeast Asia will save enough forests soon for their furry acrobats.

Glossary

acrobat (AK ro bat)—a person or animal which moves about very skillfully in high places

arboreal (are BORE ee al)—living in trees

endangered (en DANE jerd)—in danger of no longer existing; very rare

extinct (ex TINKT)—the point at which an animal species no longer exists

habitat (HAB a tat)—the kind of place in which an animal lives, such as a rain forest

predator (PRED a tor)—an animal that kills other animals for food

prey (PRAY)—an animal that is hunted by another for food

species (SPEE sheez)—within a group of closely related animals, one certain kind

INDEX